ISBN 978-0-266-59710-0
PIBN 10873110

This book is a reproduction of an important historical work. Forgotten Books uses
state-of-the-art technology to digitally reconstruct the work, preserving the original format
whilst repairing imperfections present in the aged copy. In rare cases, an imperfection in
the original, such as a blemish or missing page, may be replicated in our edition. We do,
however, repair the vast majority of imperfections successfully; any imperfections that
remain are intentionally left to preserve the state of such historical works.

For support please visit www.forgottenbooks.com

TRANSACTIONS

OF THE

American Microscopical Society

ORGANIZED 1878   INCORPORATED 1891

PUBLISHED QUARTERLY

BY THE SOCIETY

EDITED BY THE SECRETARY

VOLUME XXX

NUMBER TWO
(DOUBLE)

DECATUR, ILL.

REVIEW PRINTING & STATIONERY CO.

1911

NOTICE TO MEMBERS

Members of the American Microscopical Society are under peculiar obligation this quarter to two of our members: to Dr. J. S. Foote, who, beside furnishing the remarkable series of drawings in his article, has donated to the Society $150 toward the publication of this number; and to Mr. Edward Pennock for securing the adherence of eleven new members to the Society.

Application for entry as second-class matter at the post office at Decatur, Illinois, pending.

OFFICERS.

President: A. E. Hertzler, M.D.........................Kansas City, Mo.
Vice President: M. J. Elrod..............................Missoula, Mont.
Secretary: T. W. Galloway...................................Decatur, Ill.
Treasurer: T. L. Hankinson..............................Charleston, Ill.
Custodian: Magnus Pflaum.............................Meadville, Pa.

ELECTIVE MEMBERS OF THE EXECUTIVE COMMITTEE

R. H. Wolcott..Lincoln, Neb.
L. B. Walton...Gambier, Ohio
H. N. Ott..Buffalo, N. Y.

EX-OFFICIO MEMBERS OF THE EXECUTIVE COMMITTEE
Past Presidents still retaining membership in the Society

R. H. Ward, M.D., F.R.M.S., of Troy, N. Y.,
        at Indianapolis, Ind., 1878, and at Buffalo, N. Y., 1879.
J. D. Hyatt, of New Rochelle, N. Y.,
        at Columbus, Ohio, 1881.
Albert McCalla, Ph.D., of Chicago, Ill.
        at Chicago, Ill., 1883.
T. J. Burrill, Ph.D., of Urbana, Ill.,
        at Chautauqua, N. Y., 1886, and at Buffalo, N. Y., 1904.
Geo. E. Fell, M.D., F.R.M.S., of Buffalo, N. Y.,
        at Detroit, Mich., 1890.
Marshall D. Ewell, M.D., of Chicago, Ill.,
        at Rochester, N. Y., 1892, and at Boston, Mass., 1907.
Simon Henry Gage, B.S., of Ithaca, N. Y.,
        at Ithaca, N. Y., 1895 and 1906.
A. Clifford Mercer, M.D., F.R.M.S., of Syracuse, N. Y.,
        at Pittsburg, Pa., 1896.
A. M. Bleile, M.D., of Columbus, Ohio,
        at New York City, 1900.
C. H. Eigenmann, Ph.D., of Bloomington, Ind.,
        at Denver, Colo., 1901.
Charles E. Bessey, LL.D., of Lincoln, Neb.,
        at Pittsburg, Pa., 1902.
E. A. Birge, LL.D., of Madison, Wis.,
        at Winona Lake, Ind., 1903.
Henry B. Ward, A.M., Ph.D., of Urbana, Ill.,
        at Sandusky, Ohio, 1905.
Herbert Osborn, M.S., of Columbus, Ohio,
        at Minneapolis, Minn. 1910

The Society does not hold itself responsible for the opinions expressed by members in its published *Transactions* unless endorsed by a special vote.

TABLE OF CONTENTS

FOR VOLUME XXX, Number 2

NOTICE

The Secretary is pleased to announce to the Society that more new members have come into the Society during *four* months since the issue of Vol. 29, No. 2, than in any *twelve* months since 1891.

The officers purpose making this the most notable year's growth in the history of the American Microscopical Society. This can be done with the help of the minority of the membership now actively helping the Secretary, but the result cannot be so satisfactory as if the whole body of members were to assist.

There is no desire to make this a large, unwieldy body; but it can easily be made large enough to sustain a creditable quarterly journal. May we not confidently expect that each member will give his personal influence to secure this end?

STUDIES ON A PHOSPHORESCENT BERMUDAN ANNE-LID, ODONTOSYLLIS ENOPLA VERRILL.

BY T. W. GALLOWAY AND PAUL S. WELCH

Contributions from the Biological Department, James Millikin University, No. 7.

I. INTRODUCTION AND NATURAL HISTORY.

During the summer of 1904 the senior author, while working at the Bermuda Biological Station, had the opportunity to observe with some care two appearances of a remarkably interesting phosphorescent annelid.* The laboratory was at that time located at Hotel Frascati on the Flatts Inlet to Harrington Sound. The tide runs freely into and out of the Sound by way of the Inlet. The worms appear periodically in the waters of this inlet in considerable numbers and with striking regularity. It was reported to me that they also occur in the waters of St. George's Harbor. In all likelihood they are to be found at numerous points about the Islands.

Two appearances were observed by me at Frascati, and a third was reported to me. The first occurred July 3-7, with a maximum on the 4th; another July 29-31, with a maximum on the 30th. The latest appearance was reported to me as occurring on August 23; but details are lacking as to its duration. There is thus an interval of about 26 days between these maxima.

The full meaning of this interval is not clear. The lunar month is, however, at once suggested. This would in all probability come to be established through the tides, either alone and directly, or in connection with light variations produced by tidal variation. These agencies might possibly become operative in two ways: (1) in connection with the formation of the sexual cells, and (2) in their

*The indebtedness of the writers is hereby acknowledged for courtesies extended by the Bermuda Biological Station (supported by the Bermuda Natural History Society and Harvard University) and by Professor Edward L. Mark, its Director. The identification of the worm was made by Dr. J. Percy Moore of the Academy of Natural Sciences of Philadelphia.

release. How the effects are wrought in this case we are not in a position to say. The matter should receive more careful local study.

According to the tide tables, the tide was at its lowest for the day in the Bermudas on July 4 about 6:30 P. M., the moon being in its last quarter on July 5th. On July 27th was a spring tide; and the second recorded appearance began July 29th. On July 31 there was a near-spring tide at its lowest daily phase at 7:30 P. M. While these two appearances do not show a strict parallelism with the lunar phases, they do involve a coincidence of low tides and approaching dark.

If my informant was correct in his dates, the third appearance would occur at high tide. It is possible that the first appearance of the season, involving a release of gametes, is stimulated by low tide or by the coincidence of low tide and approaching dark; that the time necessary for renewal of reproductive bodies has been established at approximately the lunar month by a series of circumstances, internal and external; and that the release of the sexual products occurs on the first approach of dark after their maturity irrespective of the height of the tide at the time.

Parallel instances of periodicity in the formation and release of sexual bodies are numerous, in which both tide and light appear to play a part.

In addition to the monthly periodicity, there is a daily periodicity. On both occasions on which they were carefully observed they appeared each day, within fifteen minutes of the same time, just as dusk was becoming pronounced.

The display lasted from twenty to thirty minutes. Only a few appeared at first, each evening. The numbers gradually increased to a maximum, when scores might be seen at once. The display waned somewhat more rapidly than it waxed. An occasional belated specimen sometimes appeared some minutes afterward.

In a similar way, not so many individuals were seen on the first evening of each period. On the second or third night they reached a maximum, and again dwindled in numbers on following days.

Of the three appearances, that of early July was the most numerous, and that of August the least so. This suggests an annual, as well as a daily and monthly maximum; but this needs further observation.

The males and females differ considerably in size—the females often being twice as long as the males. The larger female specimens attain a length of 35 mm. Both sexes are distinctly phosphorescent:—the female with strong and more continuous glow, and the male with sharper, intermittent flashes.

In mating, the females, which are clearly swimming at the surface of the water before they begin to be phosphorescent, show first as a dim glow. Quite suddenly she becomes acutely phosphorescent, particularly in the posterior three-fourths of the body, although all the segments seem to be luminous in some degree. At this phase she swims rapidly through the water in small, luminous circles two or more inches in diameter. Around this smaller vivid circle is a halo of phosphorescence, growing dimmer peripherally. This halo of phosphorescence is possibly caused by the escaping eggs, together with whatever body fluids accompany them. At any rate the phosphorescent effect closely accompanies ovulation, and the eggs continue mildly phosphorescent for a while. The fact that the luminosity is known at no other time is further suggestive that it is produced by the material which escapes from the body cavity. If the phosphorescent glands are external, as the histology of the epidermis at least suggests, the discharge of the glands is closely correlated with ovulation.

If the male does not appear, this illumination ceases after 10 to 20 seconds. In the absence of the male the process may apparently be repeated as often as four or five times by one female, at intervals of 10 to 30 seconds. The later intervals are longer than the earlier. Usually, however, the males are sufficiently abundant to make this repetition unnecessary; and the unmated females are rare, if they are out in the open water. One can sometimes locate the drifting female between displays by the persistence of the luminosity of the eggs; but the male is unable to find her in this way.

The male appears first as a delicate glint of light, possibly as much as 10 or 15 feet from the luminous female. They do not swim at the surface, as do the females, but come obliquely up from the deeper water. They dart directly for the center of the luminous circle and they locate the female with remarkable precision, when she is in the acute stage of phosphorescence. If, however, she ceases to be actively phosphorescent before he covers the distance, he is uncertain and apparently ceases swimming, as he certainly ceases being luminous, until she becomes phosphorescent again. When her position becomes defined he quickly approaches her, and they rotate together in somewhat wider circles, scattering eggs and sperm in the water. The period is somewhat longer on the average than when the female is rotating alone; but it, too, is of short duration.

So far as could be observed, the phosphorescent display is not repeated by either individual after mating. Very shortly the worms cease to be luminous and are lost. Often they give the appearance of sinking out of sight; however, this appearance is negatived by the fact that I have caught both sexes at once by timing the current and dipping down stream, as much as six or eight feet from the point of latest visible phosphorescence. Sometimes as many as two or three males seem to take part in one mating.

The females caught and examined immediately on becoming luminous are full of eggs. Those caught after three or four displays, or after copulation, are largely empty of eggs; yet the different segments of one worm will differ widely in this particular. Eggs are often caught among the setæ and at any other points where they can be held.

Specimens in confinement after copulation may be aroused into mild phosphorescence for at least an hour.

The group of mating adaptations in this Syllid is peculiarly large and complex; and the elements entering into the precision with which the eggs and sperm are brought together are quite worth not-in. In a number of counts made of eggs captured in connection with the copulating worms, I found a range of 45-80 per cent of fertilized eggs in five batches taken at random. Considering the external fertilization of the eggs this must be considered very high. It is quite probable that this is a higher result than would be attained if the eggs and worms had been left in the sea.

The following correlated adaptations are noteworthy:

1. The concentration of the ripening and production of the ova and sperm into a few days of each month, in the worms of a

given locality; and the coincidence of the periodicity of the male and female.

2. The further concentration of these processes into 30 minutes of the twenty-four hours, and at the coming of darkness.

3. The coincidence of the luminosity with the emission of eggs and, so far as we know, its confinement to this period.

4. The repeated periods of luminosity of the female—serving as an adequate lure for the male, even though his distance may be considerable.

5. The sex-dimorphism in the character of the flashes in male and female, which may serve as recognition marks.

6. The fact that the females swim at the surface of the water at this time, while the males are beneath the surface until the former become luminous, enables the males to locate the females with greater precision. This position of the female may also make the oxidation more complete, and thus secure the increased luminosity.

7. The eyes of the male are perceptibly larger than those of the female, in spite of the fact that the females are distinctly larger than the males.

II. CLASSIFICATION.

The Syllidae, of which *Odontosyllis* is a genus, are an interesting and widely distributed family of annelids. While nothing is known of the habits and manner of life of this species of *Odontosyllis* except what is seen in this mating period, there are numerous striking features of the family recorded. They are mostly free-swimming forms; but it is believed that they dwell largely amidst the fixed vegetable and animal growths between the tide-marks, or at shallow depths, and seek their food there. Some species are known to be commensal with sponges. Certain species of *Autolytus* are reported to be parasitic on nemerteans, and other species of Polycheta. In such cases the proboscis is said to be incapable of being retracted.

To the general student the most remarkable facts about the family relate to the methods of reproduction. As in some of the naidiform Oligocheta, the non-sexual reproduction by "budding" is a common occurrence. This budding may be of the nature of transverse fission, as in the naidiform worms; or it may be lateral, sometimes even with rosettes of buds rising from the side of the body. In one species of *Autolytus* the budding may continue from these primary buds in such a way as to produce a complex, much branched stock very similar to plant growth. As will be recog-nized, this is an uncommon occurrence among animals as highly differentiated as the Syllids. In *Autolytus* and some other genera, possibly in many of the genera, a non-sexual nurse-stock gives off numerous sexual buds or zooids, which ultimately escape and mate as free-swimming worms. The embryos develop into the nurse-stock and thus a somewhat complicated alternation of generation comes about.

Malaquin (1893) diagnoses the family Syllidae as follows:

"Cephalic segments provided with 5 appendages: Namely, two palpi; two lateral and one median antennæ; and two pairs of eyes. The peristomial (post-cephalic) segment usually has two pairs of tentacle-like cirri,—some-times only one pair. The succeeding segments have feet consisting only of the setigerous lobe of the ventral division, together with a dorsal and a ven-tral cirrus. The dorsal division of the foot often develops at the time of sexual maturity. The proboscis is protrusible, and consists of two regions: (1) the anterior (pharynx) chitinous and with one or more teeth; (2) the muscular gizzard, which is a secondary development of the pharynx of the larva. Reproduction is distinguished by the appearance of secondary sexual characteristics (such as enlargement of the eyes, elongation of the antennae, development of swimming bristles and of genital glands, and often of phos-phorescent organs). The ordinary individual may thus itself become sexual by these changes (epigamy); or it may give rise to new and special buds which separate and assume the sexual characteristics (schizogamy)."

The Syllidae are divided into the following sub-families:

Syllidae.
$$\begin{cases} \text{Palps present..} \begin{cases} \text{Fused.........} \begin{cases} \text{Thruout.............Exogonea.} \\ \text{At base only..........Eusyllidae.} \end{cases} \\ \text{Separate...........................Syllidae.} \end{cases} \\ \text{Palps wanting....................................Autolytea.} \end{cases}$$

Malaquin defines the Eusyllidea, to which *Odontosyllis* belongs, thus:

"Syllidae with ventral cirrus (McIntosh says that these may be absent; they are wholly wanting in both sexes of the worm under consideration); palpi fused at the base only. Tentacular cirri filiform and cylindrical, with surface constrictions. Reproduction epigamous (direct)."

Odontosyllis was established as a genus by Claparede (1863) and is described as follows:

"Palpi short or moderately elongated; more or less separate or fused at the base. Tentacles (3) and the dorsal cirri filamentous, short—becoming longer in sexually mature zooids. Nuchal organ has a central pit. Tentacular cirri in two pairs. Ventral cirrus present [not in the sexual zooids of *O. enopla*]. Proboscis with a series of horny papillæ, the points curved backward. Ventricle (stomach) short and devoid of T-shaped cæca. Bristles with the terminal piece simple or bifid."

Verrill (1900) has described *Odontosyllis enopla* as follows:

"A large species with a dark brown, wide, short esophagus, armed with a ventral row of six stout, recurved, hook-like teeth anteriorly, besides the median dorsal tooth.

Head large, broader than long, broadly rounded in front and on the sides; posteriorly with two rounded lobes, separated by small median emargination. Eyes black, unequal, the anterior ones much larger, reniform; those of each side are so close together that they seem to be almost in contact.

Palpi shorter than the head, rather wide, thin, often wrinkled or folded in contraction, and commonly curved downward.

Tentacle tapered, rather slender, not annulated, its length about 1½ times that of the head. Antennae similar, about ⅔ as long. Tentacular cirri similar to the tentacle, the upper one rather larger and longer; the lower ones shorter; first dorsal cirrus decidedly longer and larger than the upper tentacular cirrus. Succeeding ones mostly shorter, unequal, alternately shorter and longer, tapered distally; the longer ones are equal to the breadth of the body, the shorter ones about ½ as long; those on setigerous segments 3, 4, 6, 9 are longer than the others.

The setæ are all similar, numerous, slender, short, projecting but little beyond the parapodia, with short, rather wide blades, ratio as 1:2½-3; their tips are strongly incurved and acute, with a small denticle a little distant from the end. [Fig. 16, V.] Two spiniform yellow acicula usually occur in each fascicle.

The esophagus is short and occupies about 4 segments; its margin is incurved and strongly emarginate dorsally. It bears a group of 6 [see Figs. 18, 19, 23] nearly equal, parallel, recurved hooks or teeth, which are large and strong. The conical dorsal tooth is near the margin.

The stomach is large and occupies 8 segments; it is wide, elliptical, and about twice as long as the esophagus. Its surface is covered with angular or alveolar markings, often hexagonal, so as to have a honeycomb-like appearance, but not arranged in definite rows.

Color, in formalin, is nearly white, except when containing eggs.

Length, 25mm; diameter, about 1.5mm.

One of the largest specimens has all the segments back of the gastric region filled with eggs."

III. GENERAL AND EXTERNAL MORPHOLOGY.

As in the other nereidiform worms, the body is elongated and very mobile. The length varies in the observed specimens from 19 to 35 mm. The males are distinctly smaller than the females. Just back of the head the body is almost cylindrical. The dorsi-ventral diameter grows continuously shorter from before backward, while the transverse diameter lengthens for about 40 segments, after which it too gradually diminishes.

The segments vary in number from about 110 to 130, in the twelve or fifteen specimens examined. They may be grouped in the following regions: (1) a head with 3 to 6 supecialized segments including the first setigerous segment; (2) an anterior body region of 23 or 24 segments, in which there is a gradual increase in the right-left diameter, and upon which the dorsal portion of the parapodium (notopodium) bears no setæ, although the ventral process (neuropodium) does; (3) a mid-region, comprising some 30 to 32 segments, and similar to the last segments of the preceding region but for the fact that each notopodium bears a cluster of long, swimming setæ in addition to the neuropodial setæ; (4) a posterior region, similar to the second region in that the notopodial setæ are lacking, which includes all the rest of the worm, except—(5) the specialized anal segment which bears beside the anus, a pair of elongated cirri. The variation in the number of segments in the worms is due to differences in the fourth region enumerated above. The length of the other regions is subject to very little variation.

The Head.

As is usual in the polychetes the head is well differentiated, and the problem of its segmentation is not an easy one. There is the usual pre-oral portion known as the prostomium; the mouth itself; and a region surrounding the mouth and just back of it, the peristomium. See Figs. 2, 5, 12, 13. This specialized head includes the first setigerous segment and everything in front of it. If the prostomium, as is held by Malaquin, consists of one segment only, the head of *Odontosyllis* contains four segments. If, as Pruvot thinks, the prostomial lobe represents a modification of three segments, there are six segments in the head.

If we regard the prostomium and its outgrowths as fully homologous with the structures in other regions of the body it would seem that Pruvot's contention is the sounder of the two. For the dorsal prostomial lobe bears three types of paired sensory structures: foremost of all, two ciliated palps; back of these and separated from them by a groove, the pair of lobes bearing the eyes, of which there are two on each side of the head; and, lying between these lobes, the three tentacles, one medium and two lateral (Fig. 12). It is not easy to see how this region can be regarded as segmented at all and consider it of less than three segments; but in the light of the embryonic development of this and similar worms it is questionable whether it is sound to try to homologize the divisions of this prostomial organ or even the whole organ with the segments of the body. In the larva of *Odontosyllis,* as in other Syllids, segmentation is distinctly a secondary state, superimposed on the posterior part only of trochosphere. The anterior enlarged portion seems not to share in this embryonic segmentation. In our opinion the prostomium rather represents a modification of this unsegmented anterior outgrowth of the trochosphere. It is not even morphologically a whole segment; but is rather an outgrowth and specialization of a portion of the first embryonic segment.

The peristome, on the other hand, is clearly of three segments, each bearing a pair of lateral cirri becoming progressively more dorsal. Only the third of these segments (posterior) appears, however, as a complete ring of the body. The dorsal part of this segment may protrude in the form of a flap covering the base of the lobes bearing the eyes (Figs. 1, 13, *F.*). In life this is quite distensible. A ventral projection of this same segment froms the lower lip (Fig. 5). The first and second segments show only as partial rings extending from the side of the mouth to the over-arching prostomium.

The Eyes.

The four eyes are arranged in pairs, two eyes on either side of the median line (Figs. 1, 12, 13, *Ey.*) They are mounted on protruding lobes which are capable of a certain amount of motion. The anterior eyes are somewhat larger than the posterior ones, though not markedly so. The eyes of the males are distinctly larger than those of the females. In measurements of the eyes of two males and three females, after making allowance for differences in planes of sectioning, there is a difference of 10-30 per cent in favor of the males. The opening into the pigment cup (pupil) of the anterior eyes in the normal position is directed forward and outward; that of the posterior ones, dorsally (Fig. 1, *E.*)

The cuticle which covers the front of the eye is a continuation of that which covers the exterior of the body. The lens is a spherical body lying in the cavity formed by the pigment cup (Fig. 7 *Le.*) It appears to be connected with the cuticle through the opening in the pigment cup, by a slender stalk or pedicel. After fixation the lens shows as a somewhat fibrous substance surrounding at various places roundish bodies which stain with greater intensity. It gives the appearance of a semi-fluid substance, that has been coagulated by reagents, rather than of cells. It stains readily and is rendered somewhat brittle by the usual methods of fixation and treatment.

The remaining regions of the eye—rods, pigment layer, retinal cells, and optic nerve, which form the wall of the cup of the eye are not really four distinct regions, but are continuations and differentiations of one layer of cellular elements. The wall of the cup is formed of numerous long and narrow elements *(ommatidia)* all essentially alike, with the long axes in a radial position. Each ommatidium is a highly differentiated cell (Fig. 8). The outer end of each cell narrows into a nerve fibre which enters into the optic nerve. Near this end the cell has its greatest dimensions and contains a large, conspicuous nucleus. The middle part of each of these cells is covered with a dense pigment (Fig. 8, *Pg.*) The united effect of these pigmented regions lying side by side is to produce the pigment layer which is very conspicuous and appears continuous. The inner part of the cell projects towards the lens as a clear hyaline rod. These rods form the peripheral part of the refracting mass, but are morphologically a part of the retina. The pigmented cup is continuous, except at the dorsal region where it is interrupted by a small circular aperture, the pupil, through which the pedicel of the lens passes.

Parapodia.

In the regions of the body, described above as anterior (2) and posterior (4) the dorsal ramus of the parapodium (notopodium) is specially reduced and rudimentary. It is bilobed, consisting of two short processes or tubercles. The dorsal of these is larger and bears a dorsal cirrus (Fig. 21 *Ci.*) These cirri are chiefly outgrowths of the epidermal layer of cells; but fibres from the circular layer of muscles pass to their bases. The ventral lobe of the notopodium is a smaller, pointed process, bearing no external structures in this region. In the mid-region of the body this lobe bears a conspicuous bunch of about 40 long capiliform, non-branching setæ, arranged in two parallel rows in the sac (Fig. 10, S''), and is supported by an aciculum.

The ventral ramous (neuropodium) is well developed throughout. It consists of the usual larger dorsal and smaller ventral lobes. The dorsal lobe has at its outer extremity a cleft or sac in which the setæ are imbedded. These setæ differ from the dorsal ones in being jointed, having a long stalk with a small incurved appendage (Fig. 16, $V.$). This lobe is also supported by internal acicula which reach the surface (Fig. 21, A). The setæ of both kinds sit on the basement membrane of the external epithelium and clearly arise from the modified epithelial cells lining the sac (Fig. 9, C). The ventral lobe of the neuropodium is small and does not produce the ventral cirrus usually described as characterizing the Syllidæ.

IV.   INTERNAL ANATOMY AND HISTOLOGY.

The Body-wall.

The body-wall is composed of the four usual regions; viz., (1) the cuticle, (2) the epidermis, (3) the muscular system and (4) the parietal peritoneum.

The secreted cuticle is of about the same thickness in all regions, except in a location just anterior to the peristomial flap where it is much increased in thickness (Fig. 6, E^1). It shows the following perforations: (1) the pores of the epidermal glands; (2) the points of emergence of the setæ from the body-wall; and (3) the external openings of the nephridia at the ventral basal region of each parapodium.

The cuticle takes any of the general stains, showing up peculiarly well when treated with haematoxylin. It possesses a characteristic luster which is apparent in all preparations, irrespective of the kind of treatment. Under the highest powers the homogeneous appearance gives way to a suggestion of very fine intersecting lines. The striations perpendicular to the surface are more apparent.

The epidermis is a conspicuous layer of cells in most parts of the body and is easily traceable over the exterior of the body and into the anal and oral openings. The epidermis is composed of three kinds of cells: viz., (1) the ordinary undifferentiated epithelial cells which compose the greater part of the epidermis; (2) the large regular gland cells; and (3) the "twisted" gland cells.

In all parts of the epidermis, with a few exceptions to be noted, the cells of the first class are of the usual columnar type, with conspicuous nuclei near the bases. The basement membrane is very definite in all regions, staining readily with haematoxylin. In certain quite definite regions these cells vary from the common shape. They become much more elongated near the base of the parapodium on both dorsal and ventral sides. The greatest thickening takes place, however, in the region just in front of the dorsal flap, that extends forward from the last peristomial segment, and just at the base of the brain. At this point the epidermal cells are tremendously elongated—the length in the longest being something like fifteen times the short dimension, and 6 or 8 times as long as is usually the case (Fig. 6, E'). This patch of cells has all the appearance of a sensory epithelium. There are several such patches, less striking, about the various flaps and folds of the head.

The regular glands are rather numerous, and are found in all regions of the skin. They seem to be a little more numerous on the sides of the segments above the bases of the parapodia, and on the ventral surface to the right and left of the median groove. These cells have a characteristic shape resembling a truncated cone, the larger diameter being distal. The cell-wall, nucleus, and cytoplasm are quite distinct, and the latter has a very characteristic reticulated structure which is shown in Fig. 15.

The third type of epidermal cells—the "twisted" gland cells—occur in most parts of the epidermis, and are rather numerous. They occur in the ordinary regions of the body singly or in small groups separated by one or more epithelial cells; but they are often in clusters of 4 or 5 or more on the tentacles and cirri. In general they are more numerous on the exposed parts of the body, and much less so in the depressions, though they are abundant at the lips of such grooves. They are somewhat flask-shaped, usually with irregular surface depressions that give the appearance of a spiral twist (Fig. 22). They vary considerably in size. Each cell has a distinct neck, and opens to the exterior through a pore in the cuticle. The reticulum is conspicuous throughout the cell. Usually these cells stain densely, especially at the wall. The internal reticulations are very much less stained. Often the glands show only a slight, ghostly staining, suggesting a difference of physiological state. The distribution, and apparently empty condition of many of the glands suggest that they may be the phosphorescent organs; though of this the authors have no final proof. In the dorsal region of a male specimen these glands show a somewhat different structure, as appears in Fig. 22a., *Sg*. Here the structure is of a much divided tubular sort. Their distribution and general relations, however, mark them as identical with the twisted cells.

The muscles of the body-wall present no significant departure from the condition described for other nereidiform worms. The outer circular layers give rise to acicular muscle fibers and to fan-shaped oblique fibres concerned in the motions of the parapodia. The longitudinal muscles are massed in four heavy bands—two dorsal and two ventral. They too give off tracts of fibres to the parapodia. The general relations of the muscles are very well shown in Figs. 20, 21. The muscles are unstriate.

The Alimentary Canal.

The alimentary canal is a nearly straight tube running the length of the body from the mouth to the anus. It is in no wise degenerate, as in some of the Syllids; but is functional throughout its course. Five regions may be distinguished, the last four of which are sharply differentiated from each other: (1) buccal cavity; (2) pharynx; (3) esophagus; (4) gizzard; (5) intestine.

A number of features, both embryonic and histological, seem to suggest the point of union of gizzard and intestine as the beginning of the mesenteron. There is a cuticular lining with very pronounced developments as far as the anterior end of the gizzard, and in less degree throughout its extent. It will be seen that the chief differentiations of the tract are in the stomodaeum and that the mesenteron is quite uniform in size and structure.

Mouth and Pharynx.

The first two regions, the buccal cavity and the pharynx, constitute an introvert. When withdrawn the buccal cavity occupies the peristomium, and the pharynx extends from the first setigerous segment to the seventh. When fully everted the buccal region is turned wrong side out, and the terminal opening is directly into the muscular pharynx, which is pulled forward its full length so that its posterior end is in the first or second setigerous segment. Compare Figs. 2 and 3.

The oral opening, when the introvert is withdrawn, presents a lobed margin of which the first setigerous segment furnishes the posterior ventral portion—a kind of lower lip (Fig. 5). The first and second peristomial segments form the remainder of the boundary to the mouth. The ventral floor of the buccal cavity is glandular (Fig. 2; 14, *g.*) while the sides are merely transitional from the outside to the pharynx. There is also a deep glandular fold from the dorsal wall of the cavity at the anterior margin of the pharynx (Figs. 1, 2, *O*).

The anterior part of the pharynx is very thick and muscular, but the posterior one-third, or thereabouts, is thin and pliable. It is this latter region that allows the adjustment of the phaynx to the body space when it is withdrawn into the body. This thin zone is thrown into a backward directed fold into the coelom, dorsal to the anterior end of the esophagus (Figs. 2, 3, *Ph'*.) The anterior end of the esophagus comes to lie within the lumen of the posterior part of the pharynx. The wall of the pharynx is composed of (1) the internal cuticle, (2) the epithelial lining which secretes the cuticle, (3) a thick muscular coat, and (4) the peritoneum.

The cuticle is continuous with that of the skin and similar to it in structure, but presents modifications in certain regions. It is thickened into minute conical denticles over a large portion of the pharyngeal surface (Figs. 2; 14, *dt.*). In the dorsal part the denticles begin with the mouth and extend about one-third of the length of the pharynx. In the ventral floor they begin at the posterior edge of the glandular region (Fig. 14) and cover perhaps two-thirds of the floor back of that point. The glands and the denticles do not occur together.

The epithelial layer of the pharynx is thrown into a series of longitudinal folds, somewhat irregular at first, but becoming more definite posteriorly.   One of these in the dorsal part of the pharynx has almost the definiteness of form of the typhlosole of the intestine of the earthworm (Fig. 18).   The epidermal cells of the posterior portion of the thick muscular region of the pharynx are conspicuously glandular.   In the thin region, that forms the flexure in introversion, the cells revert more nearly to the ordinary type of epithelium.

The muscular layer is the most conspicuous one in the wall of the pharynx.   It consists of a complex of circular, longitudinal, and radial or oblique fibres.   In a longitudinal section of the organ the circular fibres show themselves to be arranged on both sides of sheets, which radiate from the lumen outward not at right angles to the lumen, but in positions depending on the degree of inversion. The circular fibres appear to dominate in this organ, but it is much less easy to distinguish the various layers here than in any other part of the body.   Indeed it is easy to see that the layers are not as independent as elsewhere, and intermix to a greater degree.

The Esophagus.

The esophagus when withdrawn occupies segments 8 to 11. At its anterior margin occurs a ring of cuticular teeth which give the name to the genus.   In the ventral floor there is a row of six of these, closely appressed, uniform in size, conical in form, and curved slightly backward.   On the lateral walls near the dorsal part, and arching over so as to engage the ventral teeth, is a somewhat larger recurved tooth on either side of the lumen.   In introversion the dorsal ridge of the pharynx is pressed in between these lateral teeth (see Fig. 18, *Df.*).   It is very difficult to get satisfactory sections in the region of the teeth.

From within outward the layers in the wall of the esophagus are: The very much thickened cuticle, comprising about one-third the thickness of the whole wall; the columnar epithelium; the layer of circular muscles about the thickness of the epithelium; the longitudinal muscles, two-fifths the total thickness of the wall; and the thin peritoneum.

The cuticular layer is thicker in the esophagus than at any other part of the body. It stains densely except at the very outer margin and shows a clear striate structure, the striations running perpendicular to the surface. While the teeth are to be looked upon as special thickenings of the cuticle, there is a sharp demarcation between this layer and them (Fig. 23.)

The arrangement and character of the muscles in the region of the esophagus shows that they are not so much concerned with the mere action of the esophagus itself as with the larger problem of introversion and eversion, and of the manipulation of the teeth at the anterior end of the esophagus. It is not the purpose of this paper to trace out the course and value of the various muscular tracts; but an examination of the various figures will suggest the role of some of the muscles massed about the anterior end of the esophagus. They bring the latter organ into proper relation to the processes being initiated by the action of the pharynx, as well as furnish a point of attachment for some of the elements engaged in retraction and protrusion. The thickened cuticle of the esophagus is undoubtedly for this purpose, and for the added purpose of keeping the esophagus open and functional during the various vigorous changes of form and position of the introvert.

The Gizzard.

The gizzard or stomach is a highly developed structure immediately following the esophagus and sharply demarcated from it. It extends normally from about the 12th setigerous segment to the 21st. It practically fills the body cavity in these segments. It is of uniform size except where it tapers off rather abruptly to join the esophagus and the intestine. It is elliptical in cross-section; and the lumen is in the form of a long narrow slit, the sides of which are nearly parallel. This slit-like lumen divides the organ into symmetrical halves, whose walls are very thick except in the part opposite the edges of the lumen. Here it is not more than 1_{-5} the usual thickness. Apparently the morphological position of the slit is dorsi-ventral; but the whole organ may be rotated until it has a right left position. From a surface view, or in excentric longitudinal (tangential) section, the whole wall of the organ is seen to be crossed by a series of fine parallel lines running transverse to the organ. The band-like spaces between these lines are divided into numerous angular areas of nearly uniform shape and size. The whole presents a remarkably regular and interesting pattern (Fig. 25). Haswell (1886) has shown in his study of various species of Syllids that the wall of the gizzard is a highly complex muscular organ; and is not, as has long been supposed, a glandular one, at all. These studies confirm his conclusion in all essential particulars.

In detail, the wall of the gizzard, from within outward, shows the following regions: (1) the epithelium with its very thin secretion of cuticula; (2) a very much reduced muscular layer (Fig. 24, *mi.*) which contains internally one thickness of circular fibres and just without this a similar single thickness of longitudinal fibres; (3) a thick zone made up of muscular elements which show chiefly as radiating columns in a transverse section or in longitudinal sections radial to the organ (Figs. 21; 2, V; 24, *Col.*); (4) a thin layer of elements (Fig. 24, *mo.*) which certainly contains circular muscular fibres, and according to Haswell contains longitudinal fibres also; (5) the thin peritoneal layer.

None of these layers call for special comment with the exception of the thick muscular layer which furnishes the main body of the wall of the gizzard.

A view when the gizzard is cut longitudinally, in such a way as to display the radial elements uncut, would be illustrated by Fig. 2, V, at the left-hand side of the figure. Such a view shows three chief topographic features: (1) a series of fibrous columns passing from the thin inner muscular sheet to the outer; and (2) lying between these, open spaces somewhat shorter than the columns and tapering toward both ends; and (3) in the outer portion of these open spaces, tapering objects, thicker at the outer ends, which extend from the outer wall about one-half the way to the lumen.

Study of these objects in other sections (see Figs. 2; 24) shows the radiating columns to be muscle fibres extending the whole thickness of the wall of the gizzard. They appear as columns in the transverse section of the gizzard also (Fig. 21, 3) but do not have such well-developed spaces between them. The tapering objects extending inward between the columns present a granular appearance and are in fact annular muscle fibres cut cross-wise. They form the transverse lines that show in a tangential section, such as is seen in Fig. 25,*Cb*. They are in reality thin, wedge-like bands of muscle fibres that encircle the gizzard and separate the radial columns into successive circular zones, also shown in Fig. 26, *Cb*. Only the outer halves of the columns are thus divided, since the circular bands extend only part way to the lumen. In the cross-section of the gizzard these circular bands present their fibres in longitudinal view. They show dimly (Fig. 21, 2) in such sections outside the wavy line which appears midway the columns (Figs. 21, 24). These circular fibres are of the usual unstriate type and coalesce with the circular sheet of fibres of the inner muscular layer at the two edges of the lumen of the gizzard (Fig. 21, *Ra.*), and only there. The wavy line seen in Fig. 24 about midway on the columns seems to be the mark of the union of the inner thin edge of the circular band of fibres with the adjacent radial columns.

The histology of the radial columns has been worked out in much detail by Haswell for several Syllids. The differences between the conditions in *O. enopla* and those described by him are of minor moment. The individual fibrils run the length of the columns and are undoubtedly made up of alternate light and dark bands, such as are seen in typical cross-striate muscles. Haswell remarks that the striations found in these fibres are more strongly marked than in any crustacean or insect he had examined.

Considered as a whole, each one of these radiating columns of cross-striate muscle fibrils is made up longitudinally of two symmetrical halves hollowed out on their inner faces in such a way as almost to surround a cavity. This cavity is small and closely surrounded near the inner end of the column, as seen in cross-section (Fig. 27, f); but in cross-sections further from the lumen, the space between the two halves becomes larger, and is less completely surrounded by the muscle fibres. In the outer half of its course the halves of a given column fall on different sides of one of the circular bands of muscles referred to above. See Fig. 26, which represents a cross section of a group of the columns about two-thirds of the way toward the outer surface, and shows the circular muscles separating the triangular halves of the columns and thus throwing them into a band back to back with the nearby halves of the next row of columns. Figure 25 shows the relation of the parts just as close to the outer end of the columns as a section can be made. In this figure the outer ends of the half-columns included between two of the bands of circular fibres are shown almost to run together into a continuous zig-zag sheet. At this level the space between the circular band of unstriate fibres (Fig. 25, *Cb.*) and the adjacent zone of the radial columns (Fig. 25, *col.*) is occupied by a mass of granular nucleated protoplasm (Fig. 25, *n*). This protoplasm extends only a short distance down the columns, as appears in Fig. 28, n., which shows a section perpendicular to that in Fig. 25.

In brief summary, then, the gizzard is a complex muscular organ with an outer and inner sheet of unstriate fibres; a parallel series of thin bands of annular unstriate muscles, whose outermost fibres are in close relation to the external muscular layer, and extends inward as a band from the outer muscular sheet only about one-half the thickness of the whole wall, except that it reaches entirely to the inner muscular sheet at the angles of the lumen where the wall is thinnest and there gives off fibres which unite with it; and a series of radial columns of striate muscle fibres, each column of which is made of two trough-like halves whose concave sides are apposed in such a way as to make a conical cavity with its largest diameter at the outer end.

From the point of view of histogenesis the most interesting things about the gizzard are this mixture of striate and unstriate fibres in the single organ, and the evidence that the central cavity of the radial columns presents at the outer end a remnant of the granular protoplasm from which the muscular fibrils were differentiated. Each of these hollow columns is a simple muscular organ, in which the contractile elements are the product of the single multinucleate mass of protoplasm occupying its core and it is differentiated progressively toward the periphery, as is suggested by the last position of the nucleated protoplasm (Fig. 28 *n.*). The embryonic character of the mature fibres, and the simplicity of the relation of the fibres in the organ mark the organism as one that would well repay study upon the ontogenetic differentiation of the muscles of the gizzard.

The Intestine.

The intestine is a straight tube passing with nearly uniform character, and with gradually diminishing size, to the anal opening. It has the usual constrictions between the segments, with sacculations almost amounting to diverticula within the segments. The epithelial lining has the usual variety of columnar cells. Those in the posterior part of the tract are heavily ciliated. Both circular and longitudinal fibres may be found, but they do not form a definite or continuous sheet of tissue. Fig. 20 illustrates a cross section of the body, in the intestinal region, which includes a dissepiment. The intestine is sharply constricted; the layer of longitudinal muscle fibres appears clearly just outside the entoderm; and a thick zone of circular fibres outside this shows a clear anastomosis with the circular fibres of the body wall in both dorsal and ventral regions. The distribution of oblique muscle fibres in the dissepiment is also shown—the origin being in the ventral region on either side the nerve cord and passing out fan-wise to the muscular layers of the dorsolateral walls. At least some of these fibres run into the circular muscular sheet in being inserted.

The only noteworthy differentiations in its length are the shallow proctodeum, and the valvular enlargement of the intestine at its junction with the gizzard (Fig. 4, *v.*). At the beginning of the intestine there is a special annular or sphincter group of muscle fibres (Fig. 4, *S*).

The Circulatory System.

The blood vessels agree with those described as characterizing the Syllids generally, and call for little special discussion. The dorsal vessel is small and thick-walled as compared with the ventral. The dorsal vessel is imbedded in the wall of the intestine in an interesting way thru a part of its course—the peritoneal membrane inclosing both in one sheath (Figs. 17; 29). Segmental vessels arise from both the dorsal and ventral longitudinal vessels and run outward on the anterior face of each dissepiment.

The Excretory System.

The nephridia are quite large organs in this worm in comparison with the size of the body. This is due in the female to the liberal glandular portion. The nephridia apparently occur in all the setigerous segments of the body beginning with the very first—which is also the last segment of the peristomium.

In the male the tubular ciliated portion of the nephridium is much larger than in the female, and the glandular part appears less massive (Fig. 11, *N.*, *G.*). In the sections of a female, whose cavity was filled with eggs in such a way as to indicate that ovulation had not occurred, the tubular part of the nephridia was so compressed and displaced as to preclude the possibility that they could be used for the discharge of eggs, at least at the beginning of ovulation. In the male on the contrary there is abundant evidence that these organs are in an hypertrophied condition, and probably they function as vasa deferentia. They have large lumens, and protrude well into the middle of the body.

Genital Products.

As in other polychetes the eggs and sperm are produced by proliferation from the coelomic peritoneum. They are found in all the full-sized segments of the body, beginning about the 19th or 20th setigerous segment. Ripe eggs were found in sections two or three segments anterior to the union of gizzard and intestine. Fig. 30 shows a cross-section of a female which has not deposited any of her eggs. In this case the eggs are so numerous that the segment is stretched until the body wall is very thin. The eggs occupy every available cavity. As has been indicated it appears to the writers that the ova could not use the nephridia in this case, but must escape by rupture of the body wall.

BIBLIOGRAPHY

ANDREWS, A. E.
   1892. The eyes of Polychaetous Annelids. Journal of Morphology, 7:169.

BENHAM, W. B.
   1896. Polychaete Worms. Cambridge Nat. Hist. Vol. 2, Chs. 9-13.

CLAPAREDE, A. R. E.
   1863. Beobachtungen über Anatomie und Entwicklungsgeschichte wirbelloser Theire, pp. 38-48.

EHLERS, E.
   1864. Die Borstenwürmer. Erste Abt. pp. 203-232.

GALLOWAY, T. W.
   1908. A Case of Phosphorescence as a Mating Adaptation. Sch. Sci. and Math. May, 1908.

HASWELL, WM. A.
   1886. On the Structure of the So-called Glandular Ventricle (Drüsenmagen) of Syllis. Quart. Journ. Micr. Sci. 26:471.
   1889. A Comparative Study of Striate Muscles. Quart. Journ. Micr. Sci. 30:31.

LANGERHANS, P.
   1879. Die Wurmfauna von Madeira. Zeitschrift für wiss. Zoologie. Bd. 32:554.

MALAQUIN, ALPHONSE.
   1893. Recherches sur les Syllidiens. Lille.

McINTOSH, WM. C.
   1873. A Monograph of British Annelids. Ray Society. Three Vols.

MOORE, J. PERCY.
   1909. Polychaeteous Annelids dredged off the Coast of Southern California. Proc. Phila. Acad. Nat. Sci. 61:321-351.

ST. JOSEPH, A. DE.
   1886. Les Annèlides Polychètes des Cotes de France. Ann. d. Sci. Natur. Zool. Ser. 9:vol. 3; pp. 145-260.

VERRILL, A. E.
   1900. Additions to the Turbellaria, Nemertina, and Annelida of the Bermudas. Trans. Conn. Acad. Sci. 10:pt. 2:627-8.

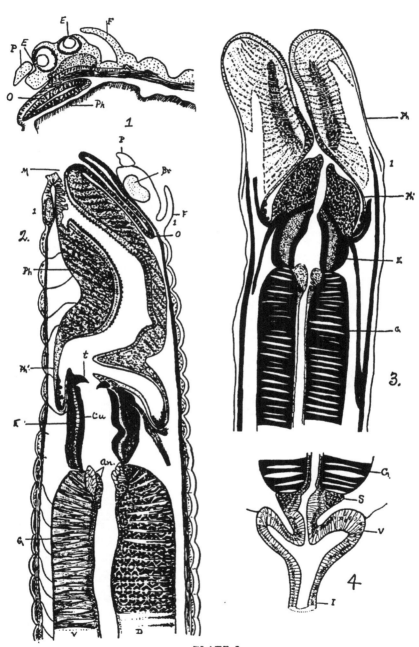

PLATE I

EXPLANATION OF PLATES

Plate I

Fig. 1. A para-sagittal section of the dorsal part of the head. Diagrammatic. *E*, the pupils of the eyes (see also Fig. 7); *F*, flap of peristomium (see also Fig. 6); *O*, a glandular fold at the dorsal anterior edge of the pharynx (see also Fig. 2); *P*, palp; *Ph*, pharynx.

Fig. 2. A sagittal section of the head and anterior segments, with the introvert withdrawn. *1*, the last peristomial (first setigerous) segment; *an*, an annular organ in the cardiac opening of the gizzard, consisting of modified epithelium and circular muscle fibres; *Br.* brain; *Cu*, cuticula of the esophagus *(E)*; *F*, a flap projecting forward from the 3rd peristomial segment (see also Figs. 1 and 6); *G*, gizzard, in which the dorsal *(D)* and ventral *(V)* walls are cut in different relation to the elements composing the wall; *M.* mouth; *O*, a thin-walled glandular pocket at the anterior, dorsal edge of the pharynx; *P*, palp; *Ph*, pharynx, with thick muscular wall; *Ph'*, thin-walled, flexible portion of pharynx; *t*, cuticular teeth in anterior margin of esophagus. ×40.

Fig. 3. A frontal section thru head and anterior segments, with the introvert protruded. Lettering as in Fig. 2. ×40.

Fig. 4. Longitudinal section at junction of gizzard and intestine. *G*, gizzard; *I*, intestine; *S*, sphincter muscle at beginning of intestine; *V*, valvular enlargement at anterior end of intestine. ×40.

Plate II

Fig. 5. Ventral view of the head; *l,* lateral tentacle of the prostomium; *m,* median tentacle; *P,* palpus; 1, 2, 3, the peristomial tentacles or cirri. The lower lip is shown to be an anterior projection of the first setigerous segment. ×15.

Fig. 6. Longitudinal section thru dorsal body-wall immediately in front of 3rd peristomial (first setigerous) segment, B, brain; E, epithelium; E', the thickened sensory epithelium in front of the peristomial flap *(F).* ×300.

Fig. 7. Eye. A median section of anterior eye thru pupil. *Cu,* cuticula; *E,* epithelium; *Le,* lens; *o. n.,* optic nerve; *Pe,* pedicel of the lens; *Pg,* pigment layer of the retina; *r,* rods, or layer formed by inner ends of retinal cells; *R,* retinal cells, deeper portion containing nuclei. ×330.

Fig. 8. Ommatidium. *C,* the nucleated part of the cell; *F,* fibre; *Pg,* pigment, which collectively makes the pigment cup of the eye; *r,* the rod or inner end of cell. ×500.

Fig. 9. Setae-sac of dorsal bristles, cut in the long axis of the setae. *A,* aciculum; *c,* bristle cell, from which the bristles arise; *m,* basement mebrane; *s,* seta (see also Fig. 16). ×125.

Fig. 10. Cross-section of two parapodia showing both the dorsal and ventral bristle sacs cut across the long axis of the setae. *A,* acicula (the smaller light objects are the setae); *Ci,* dorsal cirrus; *n,* nephridium; *S',* ventral seta-sac; *S",* dorsal sac, with the swimming bristles in two parallel rows. The darker dots show muscular elements. ×50.

Fig. 11. Cross-section of a segment in the mid-region of the body of a male. *G,* glandular part of nephridium; *N,* the tubular portion of nephridium, much enlarged and used as sperm-ducts; *T,* the twisted gland cells (see Fig. 22) numerous in the dorsal body wall. ×50.

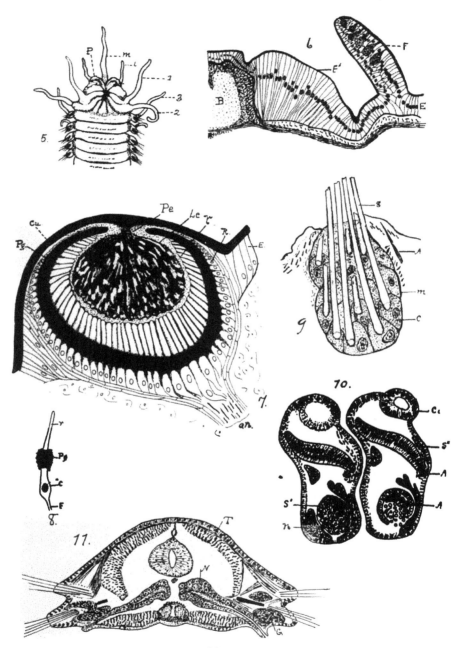

PLATE II

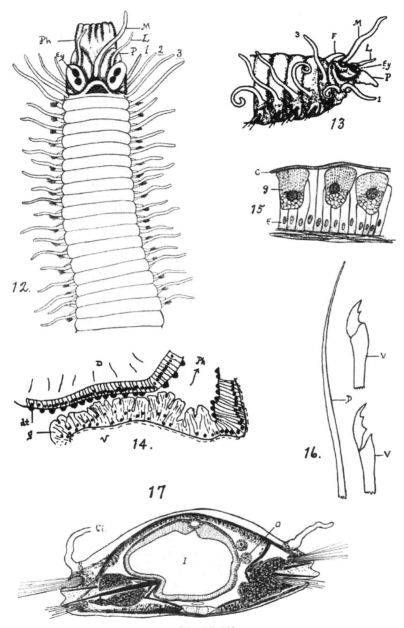

PLATE III

Plate III

Fig. 12. Dorsal view of head and anterior body segments. *Ey,* eyes, occurring in pairs on flaps; *L,* lateral tentacles of the prostomium; *M,* median tentacle; *P,* palpus; *Ph,* introvert protruded; 1, 2, 3, the tentacles of the three peristomial segments. ×15.

Fig. 13. Lateral view of the head. Lettering as in Fig. 12. ×15.

Fig. 14. Sagittal section thru buccal cavity, showing beginning of the pharynx. *D,* dorsal; *V,* ventral, *dt,* conical denticles of cuticula in wall of pharynx; *g,* a patch of glandular cells in floor of mouth; *Ph,* lumen of pharyx.

Fig. 15. Gland cells of epidermis. *C,* cuticula; *E,* epithelium; *g,* gland cell showing characteristic inner reticulated protoplasm and outer striated protoplasm. ×600.

Fig. 16. Setae. *D,* the dorsal, capilliform, swimming setae of the mid-body region; *V,* the ventral jointed setae, occurring the whole length of the body.

Fig. 17. Cross-section of the body in the mid-region of body. *Ci,* dorsal cirrus; *I,* lumen of intestine; *O,* ova in body cavity. ×50.

Plate IV

Fig. 18. Cross-section of pharynx and esophagus, with introvert withdrawn. E, muscular wall of the esophagus, showing the lateral mass of muscular fibres connected with the tooth apparatus; *ep*, epithelial layer that secretes the ventral teeth; Df, a longitudinal, dorsal fold of the pharynx that invades the front of the esophagus in introversion; *Ph*, muscular wall of the pharynx. ×65.

Fig. 19. Similar cross-section a little anterior to Fig. 18. ×65.

Fig. 20. Cross-section in mid-region of body, including a dissepiment (*Di*). 1, dorsal ramus of notopodium, bearing dorsal cirrus (*Ci*); 2, ventral ramus of notopodium, bearing swimming bristles supported by aciculum; 3, dorsal ramus of neuropodium, bearing bristles and acicula; 4, ventral ramus of neuropodium with no appendages. ×50.

Fig. 21. Cross-section of body thru the gizzard. *A*, aciculum; *Ci*, dorsal cirrus; m^1, layer of circular muscles of body wall; m^2, longitudinal muscles; *Ra*, raphe where there is an anastomosis of the fibres of the annular band of muscle fibres (2) with the innermost layer of the gizzard. The numerals indicate three successive zones of elements in the wall: 1, the peritoneum and thin outer muscle layer; 2, annular bands of fibres, much flattened in the direction of the long axis of the animal suggested by the broken lines; 3, muscular columns which make up the bulk of the wall of the organ. Within this is a thin layer consisting of muscles, epithelium, and cuticula. ×65.

Fig. 22. "Twisted" glands (phosphorescent?) from epidermis of female. *Po*, pore in cuticula; *Sg*, the gland with constrictions. ×680.

Fig. 22a. Similar glands in the dorsal epithelium of a male specimen. ×680.

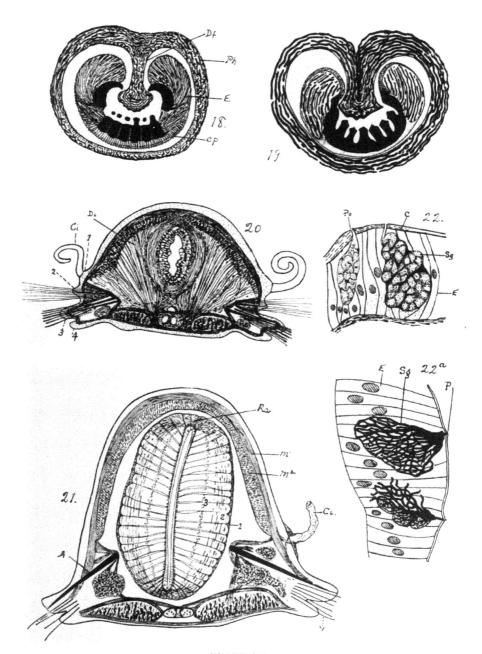

PLATE IV

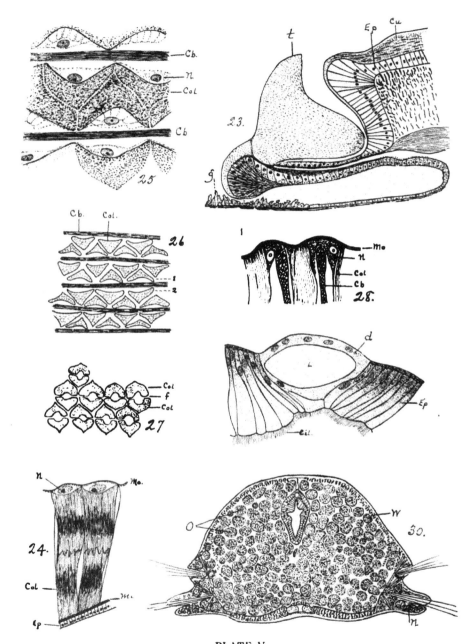

PLATE V

Plate V

Fig. 23. Detail of ventral floor of esophagus, longitudinal section, in region of teeth. *Cu*, cuticula lining esophagus; *Ep*, epithelial layer; *g*, glandular epithelium in the thin portion of pharynx; *t*, one of the ventral teeth.

Fig. 24. Two muscle columns of the gizzard, taken from transverse section of that organ (see Fig. 21). Lettering in figures 24-28: *cb*, "circular band," a thin sheet of muscle whose fibres encircle the organ; *col*, the radial muscular columns whose fibres radiate from the lumen; *f*, a fissure separating the halves of one column; *mi*, thin inner muscular layer; *mo*, a thin outer muscular layer; *n*, nucleated protoplasm between the outer ends of the half-columns.

Fig. 25. Section perpendicular to 24, transverse to the columns and tangential to gizzard, just as near the outer surface as possible. ×320.

Fig. 26. Section transverse to the columns (parallel with 25), at a deeper level. ×160.

Fig. 27. Section parallel with 24 and 25, transverse to the columns at their inner end). ×160.

Fig. 28. Longitudinal section of columns (outer end) at right angle to Fig. 24, showing them as they appear in longitudinal section of the gizzard. ×160.

Fig. 29. Dorsal wall of intestinal tract showing relation of dorsal blood vessel to it. *cil*, cilia; *d*, dorsal blood vessel; *ep*, epithelial lining of digestive tract; *L*, lumen of blood vessel. ×360.

Fig. 30. Cross-section from mid-body region of a female in which ovulation has not commenced. *N*, glandular part of nephridium; *O*, ova; *W*, the much stretched and thinned dermo-muscular wall. ×40.

OF THE

AMERICAN MICROSCOPICA SOCIETY

PUBLISHED QUARTERLY BY THE

MERICAN MICROSCOPICAL SOCI
DECATUR, ILLINOIS

SUBSCRIPTION $3.00   SINGLE

DOUBLE NUMBERS $2.00

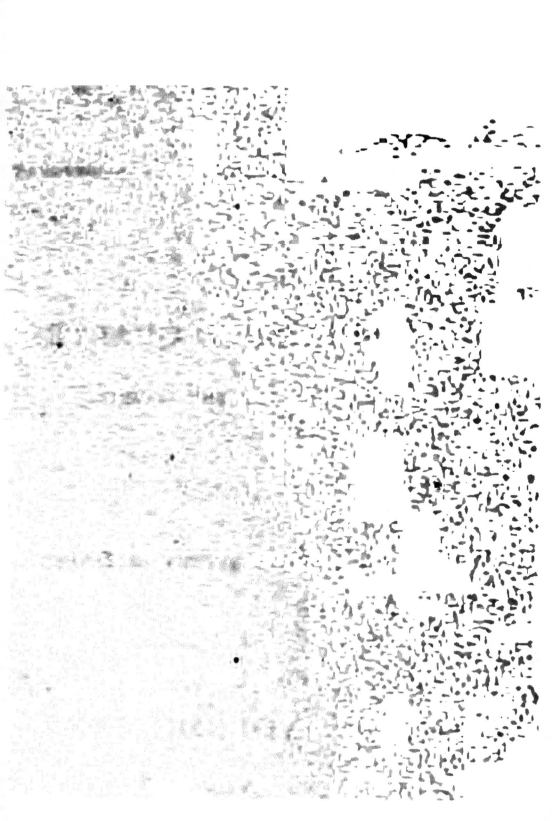

TRANSACTIONS

OF THE

American Microscopical Society

ORGANIZED 1878   INCORPORATED 1891

PUBLISHED QUARTERLY

BY THE SOCIETY

EDITED BY THE SECRETARY

VOLUME XXX

NUMBER THREE

DECATUR, ILL.
REVIEW PRINTING & STATIONERY CO.

1911

NOTICE TO MEMBERS

The twenty-year period for which the American Microscopical Society was originally incorporated expired in August, 1911. Some months previous to this, the Secretary, authorized by the legal proportion of the Executive Committee of the Society, including the present officers, had this incorporation made perpetual.

Application for entry as second-class matter at the post office at Decatur, Illinois, pending.

TABLE OF CONTENTS

TRANSACTIONS
OF
American Microscopical Society
(Published in Quarterly Installments)

| Vol. XXX | APRIL 1911 | No. 2 |

THE COMPARATIVE HISTOLOGY OF FEMORAL BONES.

By J. S. FOOTE, M. D.

It is interesting to notice the diversity of structure present in the femurs of various animals. We have been led to think that bone is bone; that long bones, and especially similar bones wherever they occurr, have the same structure because they are used for the same purposes. This is not true. Microscopical examinations of entire sections of femurs of forty-three different animals show that the structures and their arrangements are not at all the same. They vary in different animals and even in the walls of a bone of any one animal. The anterior wall is unlike the posterior, the outer wall unlike the inner. Nor are the lineal portions of a bone the same. If a long bone, like the femur, is divided into thirds, each third is found to differ from another third. Therefore, a correct knowledge of the histology of bone cannot be obtained from pre-pared sections of small pieces sawed from the walls of bones. Entire cross sections must be made in each case. Furthermore, a drawing and description of one long bone will not answer for another.

In a general way femurs answer a common purpose, but in specific ways they differ considerably. The various habits of animals, the complex muscular stresses to which their femurs are exposed and the variations in position relative to the bodies which they support demand corresponding structural arrangements peculiar to the animals possessing them. Thus the femurs of swimmers, flyers, scratchers, climbers, crawlers, runners and jumpers differ from each other very decidedly. Bone exhibits a very ready structural response to functional demands.

We are accustomed to one type of bone, viz: the Haversian system type. It is the familiar type described and plated in books. It is human and will not answer as a formula of the general bone structure of animals. In most cases a small piece of a long human bone is described. In some instances entire cross sections of small bones, as the fibula, radius, metatarsal or metacarpal, are given. This type, in its pure form, appears to be peculiar to man, as none of the other femurs examined show it. The completely developed Haversian system evidently belongs to the higher animals.

Entire cross sections of the femurs of the following forty-three animals have been examined as they were received. They were not selected:

LIST OF BONES EXAMINED

AMPHIBIA.
Frog (4 specimens). Pl. A; figs. 1-3; Pl. I; figs. 1, 2.

REPTILIA.
Alligator, Pl. I, fig. 4.
Snapping Turtle, Pl. I; figs. 5, 6.

AVES.
Order *Steganopodes*
Pelican, Pl. I, fig. 7.

Order *Anseres*
Mallard Duck, Pl. II; fig. 8.
Wild Goose, Pl. II; fig. 9.

Order *Striges*
Owl, Pl. II, fig. 10.

Order *Accipitres*
Eagle, Pl. II; fig. 11.
Hawk, Pl. II; fig. 12.

Order *Gallinae*
Grouse, Pl. II; fig. 13.
Chicken, Pl. III; fig. 14.
Prairie Chicken, Pl. III; fig. 15.
Domestic Turkey, Pl. III; fig. 16.
Wild Turkey, Pl. III; fig. 17.
Peahen, Pl. III; fig. 18.

Order *Picariae*
Yellow Hammer, Pl. III; fig. 19.

Order *Passeres*
Crow, Pl. IV; fig. 20.
Jay, Pl. IV; fig. 21.

MAMMALIA
Marsupials
Opossum, Pl. IV; fig. 22
Placentals
Order *Rodentia*
Musk Rat, Pl. IV; fig. 24.
Rat, Pl. IV; fig. 25.
Rabbit, Pl. IV; fig. 26.
Squirrel, Pl. IX; fig. 50.
Woodchuck, Pl. V; fig. 27.
Prairie Dog, Pl. V; fig. 28.

Order *Carnivora*
Skunk, Pl. V; fig. 29.
Raccoon, Pl. V; fig. 30, 32.
Mink, Pl. V; fig. 31.
Weasel, Pl. VI; fig. 33.
Wild Cat, Pl. VI; fig. 34.
Cat, Pl. VI; fig. 35.
Gray Fox, Pl. VI; fig. 36.
Wolf, Pl. VI; fig. 37.
Dog, Pl. VI; fig. 38.

Order *Ungulata*
Elk, Pl. VII; fig. 39.
Deer, Pl. VII; fig. 40.
Ox, Pl. VIII; fig. 43.
Sheep, Pl. VIII; fig. 44.
Goat, Pl. IX; fig. 51.
Horse, Pl. VIII; fig. 41.
Pig, Pl. VIII; fig. 42.

Order *Primates*
Monkey, Pl. IX; fig. 52.
Man, Pl. IX; fig. 53.

All sections are taken from the middle of the femurs, ground to proper thinness, mounted in hard balsam and examined with the same objectives, oculars and tube lengths of a Zeiss microscope. Drawings are then made. As some of the femurs are very large and some are very small, all drawings are made from the viewpoint of clearness rather than from actual sizes.

The drawings do not make any effort, therefore, to give the exact number of Haversian systems, laminae, lamellae, etc. The aim has been merely to show the relative positions, arrangements, proportions and developments of these structures. The horizontal line in connection with each femur gives the natural diameter of the bone.

The following general outline is followed in all examinations: (1) antero-posterior diameter of the bone; (2) lateral diameter of the bone; (3) antero-posterior diameter of the medullary canal; (4) latera ldiameter of the medullary canal; (5) medullary canal, full or empty; (6) trabeculae of bone, present or absent; (7) cancellous bone, present or absent; (8) compact bone; (9) hardness or density; (10) character of external and internal circumferential lamellae; (11) arrangement, development and character of the Haversian systems; (12) laminae—concentric or oblique, position of; (13) lacunae, character of; (14) canaliculi, character of; (15) type of structure.

DETAILED DISCUSSION.
Femurs of the Bull Frog.

The femurs of four frogs are examined, the first unusually large, the second of medium size and the third and fourth small. The first femur is 3.5 mm. x 4.5 mm., the second 2.5 mm. x 3 mm., the third and fourth 1 x 1.3 mm.

They show different developments of the same type of bone (Pl. A, Figs. 1, 2, 3; Pl. I, Figs. 1-3).

Femur of the First Frog.
Pl. A, Fig. 1.

Antero-posterior diameter of bone, 3.5 mm.; lateral, 4.5 mm.
Antero-posterior diameter of medullary canal, 1 mm.; lateral, 2 mm.

The medullary canal is full. There are no trabeculæ and no cancellous bone. The bone is soft.

lamellar type, though of different developments. In Fig. 1, the radiating canals, with poorly developed intervening lamellae, indicate a low stage of development. In Fig. 2 half of the canals have disappeared and better developed lamellae are formed. In Fig. 3 all of the canals have disappeared and the whole bone consists of concentric lamellae.

Fractured and Repaired Femur of *the Frog.*
Pl. I, Fig. 2, 3.

One of the femurs (Fig. 2) had been fractured about the middle of the shaft and repaired. The ends of the bone had slipped by each other and new bone had formed around the fragments. In the section, which was taken from the middle of the new bone, two cuts of the femur appear situated eccentrically. The upper fragment, proximal, shows cell growths bursting through the wall of the bone (Pl. I, Fig. 2, A. B.). In the lower fragment, distal, no cell outbursts appear.

Around the two fragments and extending between them is a formation of cancellous bone which is the new bone of repair. The cancellous bone resembles large Haversian systems, although there are no Haversian systems in the femur. This fact suggests a genetic relationship between cancellous bone and Haversian systems, and also indicates that bone repairs are made by cancellous bone. Evidently the lamellar type is the simplest type of bone structure.

Femur of *the Alligator.*
Pl. I, Fig. 4.

Antero-posterior diameter of the bone, 15 mm.; lateral, 17 mm.

Antero-posterior diameter of medullary canal, 5.5 mm.; lateral, 6 mm.

The medullary canal is full. No trabeculæ. Very little cancellous bone. The bone is hard.

Structure. A thin cross section of this femur held up to the light presents a ringed appearance like that of a cross section of the trunk of a tree.

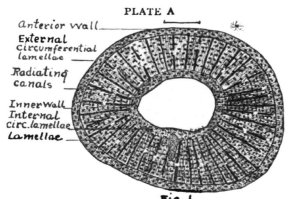

anterior wall

External
circumferential
lamellae

Radiating
canals

Inner Wall
Internal
circ. lamellae
Lamellae

Fig. 1.
I...mur of a large Frog showing
lamellae interrupted by radiating
Canals. Low type of structure.

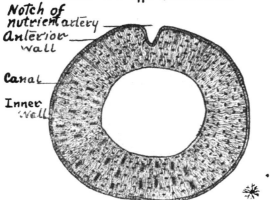

Notch of
nutrient artery

Anterior
wall

Canal

Inner
wall

Fig. 2
Femur of Frog, showing a struct-
intermediate between figs. 1 and 3.

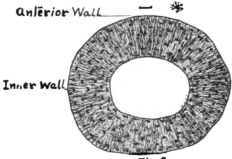

anterior Wall

Inner Wall

Fig. 3.
Femur of Frog, showing
lamellar structure.

Under the microscope the following concentric rings are found, beginning with the circumference:

1. A wide ring composed of irregularly-shaped canals enclosed within a network of canaliculi, radiating from long and oval lacunae, embedded in bone substance. Lamellae do not appear. The ring has the appearance of very incomplete Haversian systems. The lacunae are small or large and their canaliculi are long, branching and bushy.

· 2. A second narrow, laminar ring composed of two or three lamellae between which are long narrow lacunae, with long, straight canaliculi. The lamellae are well developed.

3. A second ring of incomplete Haversian systems, narrower than the first, but of similar construction.

4. A second narrow lamina, like the first.

5. A third ring of incomplete Haversian systems, like the other two, excepting that it is a little denser.

6. A third lamina, like the two first described.

7. A fourth wide ring of incomplete Haversian systems like the others.

8. A fourth narrow ring of internal circumferential lamellae, three or four in number, interrupted by a little cancellous structure in the inner and outer walls. It may be noticed that the bone has four concentric laminae alternating with four rings of incomplete Haversian systems, which is practically the same structure as found in the femur of the turtle. The bone is hard.

Its peculiar features are the absence of complete Haversian systems, its uniform concentric ringed structure, its laminar development and the presence of incomplete Haversian systems.

Femur of the Snapping Turtle.
Pl. I, Figs. 5, 6.

Antero-posterior diameter, 8 mm.; lateral diameter, 8.5 mm.
Antero-posterior diameter of medullary canal, 1 mm.; lateral diameter, 1 mm.
Medullary canal is full. No trabeculæ.

Structure. The walls of the shaft are very thick, proportionately, and the medullary canal is very small. The femur is nearly solid. Around the medullary canal is a zone of cancellous bone. Around this is a thick zone of compact bone. The bone is hard. This variation in the relative diameters of the shaft and medullary canal is in marked contrast with the measurements of other femurs. The type of structure is mixed and incomplete. The bone presents quite a complicated arrangement of its structural units, lamellae, laminae and Haversian systems. The following structures appear, beginning with the outer boundary and proceeding toward the medullary canal:

1. A clear peripheral lamella of bone containing only a few irregularly-shaped lacunae, with few canaliculi.

2. A complete concentric lamina of bone composed of four or five lamellae closely united. The lacunae are oval, round or long and narrow and their canaliculi are numerous and bushy.

3. A wide ring of incomplete Haversian systems. These systems are composed of oval or round lacunae arranged around rather large Haversian canals. The canaliculi are radiating in arrangement. The lamellae of the systems are not distinctly marked. The canals are large. The systems present an appearance of incompleteness.

4. A second concentric lamina composed of three or four lamellae of bone with long narrow lacunae and bushy canaliculi.

5. A second wide ring of Haversian systems similar in all respects to the first.

6. A third concentric lamina of three or four lamellae similar to the others described.

7. A third ring of Haversian systems, narrower than the others, otherwise similar.

8. A fourth concentric lamina similar to the others described.

9. A wide zone of cancellous bone surrounding the medullary canal.

Thus the femur has four concentric laminae alternating with three rings of incomplete Haversian systems. The laminae appear to be more completely developed than the systems. All of the laminae and rings of Haversian systems, at one point of the section, bend inward from the external surface to the cancellous central bone.

Femur of the Pelican.

Pl. I, Fig. 7.

Antero-posterior diameter of the bone, 11.5 mm.; lateral, 12 mm.
Antero-posterior diameter of the medullary canal, 9.5 mm.; lateral, 10 mm.
The medullary canal is full. There are no trabeculae. The bone is of medium hardness. No cancellous bone.

Structure. External circumferential lamellae, four to ten in number, surround the bone, excepting at the anterior and outer posterior ridges. The lamellae are well developed, their lacunae are long and narrow and their canaliculi are long and branching. At the anterior and posterior ridges the lamellae are interrupted by tendon insertions.

The central ring is composed of Haversian systems in different stages of development. Underneath the external lamellae is a narrow ring of well-developed Haversian systems.

The main portion of the bone is composed of oval and round lacunae, with short, bushy canaliculi, forming a delicate network within the bone substance. There are no lemellae, laminae or Haversian systems. Wide branching canals extend from the medullary canal outward and cross the bone in all directions, forming a coarse network. This portion of the bone resembles reptilian bone (Pl. I, Fig. 4, 5, 6).

The posterior wall consists of rather indistinct whorls of lacunae and their reticular canaliculi bordering on the medullary canal. In some places half of a system forms the boundary line. The outlines of the Haversian systems are more clearly marked in the internal than in the external half of the wall. The lacunae are long and narrow. The internal circumferential lamellae, two or three in number, surround the medullary canal, excepting in the posterior wall. They form an extremely narrow boundary of the medullary canal. The lamellae are only partly developed. The lacunae are oval and the canaliculi are short and bushy.

The section shows a thickening a little to the inner side of the anterior mid line. The external surface has tendon insertions, extending through the external circumferential lamellae. The canals of this region have a transverse direction. There is another thickening in the outer wall formed of crude Haversian systems and two slight thickenings in the posterior wall formed in a like manner. The type is Haversian system, undeveloped. The peculiar feature of the bone is its low stage of development.

Femur of the Mallard Duck.
Pl. II, Fig. 8.

Antero-posterior diameter of the bone, 4.6 mm.; lateral, 6.3 mm. Antero-posterior diameter of medullary canal, 3.5 mm.; lateral, 5.5 mm.

Medullary canal empty. No trabeculæ and no cancellous bone. The bone differs from that of the wild goose in that it does not show the division into outer and inner rings of Haversian systems and laminæ. It is of medium hardness.

Structure. 1. External circumferential lamellae, four to six in number, with long, narrow lacunae and branching canaliculi.

2. A wide ring of various combinations of irregularly-shaped Haversian systems and short laminae. No plan of arrangement is evident. The systems appear to be a laminar formation doubled or rolled into crude forms. Their lacunae are rather few and their canaliculi are short and bushy. Their lamellae are clearly marked. They run in all directions, as may be seen from their cross sections. The laminae are short (inter-Haversian), and the canals are numerous and frequently intersecting. Their lacunae are long or oval and their canaliculi are bushy.

3. Internal circumferential lamellae, three or four in number, with long lacunae and branching canaliculi. Along the posterior ridge and surface the Haversian systems are more numerous and better developed.

The bone is of the mixed type, but incompletely developed.

Femur of the Wild Goose.
Pl. II, Fig. 9.

Antero-posterior diameter of the bone, 9 mm.; lateral, 8.5 mm. Antero-posterior diameter of medullary canal, 7 mm.; lateral, 7 mm.

PLATE I

Cancell

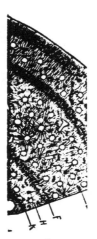

Fig. 6
ig.5 enlarge

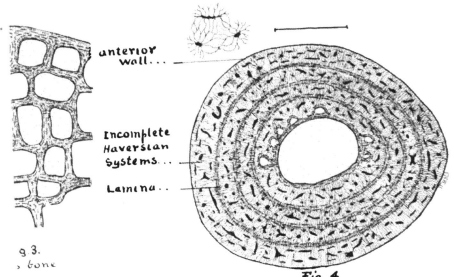

anterior
wall...

Incomplete
Haversian
Systems...

Lamina..

g 3.
, bone

Fig. 4
Femur of an Alligator showing
Concentric rings of very incomplete
Haversian Systems and laminae

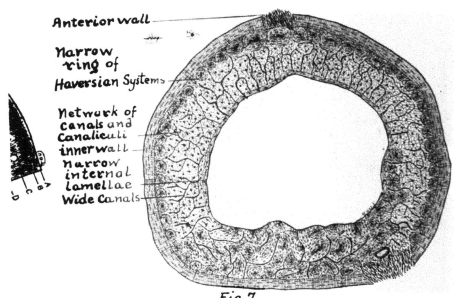

Anterior wall

Narrow
ring of
Haversian Systems

Network of
canals and
Canaliculi
inner wall
Narrow
internal
lamellae
Wide Canals

Fig. 7
Femur of a Pelican-showing a low.
stage of development.

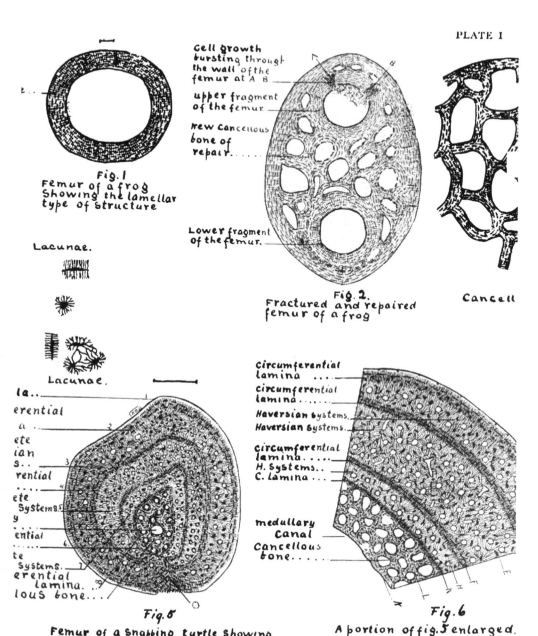

PLATE I

Fig. 1
Femur of a frog
Showing the lamellar
type of structure

Lacunae.

Lacunae.

Cell growth
bursting through
the wall of the
femur at A B.

upper fragment
of the femur.

new cancellous
bone of
repair.

Lower fragment
of the femur.

Fig. 2.
Fractured and repaired
femur of a frog

Cancell

la..
erential
a .
ete
ian
s..
rential
ete
Systems.
y
ential
te
Systems.
erential
lamina. ..
lous bone...

Fig. 5
Femur of a snapping turtle showing
alternating Haversian Systems
and circumferential laminae and
small medullary canal. medullary
Canal is very small.

circumferential
lamina ...
circumferential
lamina.......
Haversian Systems.
Haversian Systems.
circumferential
lamina. ....
H. Systems..
C. lamina ...

medullary
Canal
Cancellous
bone.....

Fig. 6.
A portion of fig. 5 enlarged.

Medullary canal is full. No trabeculæ and no cancellous bone. The bone is of medium hardness. The medullary canal is large and the walls of bone are thin.

Structure. The external circumferential lamellae are well developed and vary from four to ten in number. Their lacunae are long and narrow and their canaliculi are rather few in number. At the posterior ridge are found many Haversian systems incompletely developed. Their lacunae are round or oval and their canaliculi are few. The anterior wall of the bone is composed of incomplete Haversian systems occupying the whole thickness of the wall between the external and internal circumferential lamellae. The remainder of the bone, about four-fifths of the whole, shows quite different structures and arrangements in the two walls, outer and inner. The outer wall is composed of irregularly-shaped laminae and undeveloped Haversian systems. The laminae are confined to the inner half of the wall, while the Haversian systems occupy the outer half, situated under the external circumferential lamellae.

The two halves are well marked and distinct from each other. The lacunae of the systems and laminae are oval and their canaliculi are few and short. The Haversian canals of the laminae are irregular and branching. The inner wall of the shaft is composed of an internal band of well-developed laminae and an outer band of rather poorly-developed Haversian systems. The band of Haversian systems is just under the external circumferential lamellae, the band of laminae is just outside of the internal circumferential lamellae. The laminae are better developed than the systems. The peculiar structural feature of the whole wall of the bone is its division into two equal concentric rings, one of Haversian systems and the other of laminae. The internal circumferential lamellae, from three or four to ten or twelve in number, are well developed. On'the left side of the bone quite large canals extend from the medullary canal through the laminae to the Haversian canals between the laminae. The type is mixed.

The bone has the following rings beginning with the outside:

1. External circumferential lamellae.
2. Irregular laminae and Haversian systems.

3.   Well-developed laminae.

4.   Internal circumferential lamellae.

Indistinct arrangements of similar structures are present in the femurs of several other birds.

Femur of the Large Horned Owl.
Pl. II, Fig. 10.

Antero-posterior diameter of the bone, 8 mm.; lateral, 7.5 mm.

Antero-posterior diameter of medullary canal, 6 mm.; lateral, 6 mm.

Medullary canal is full.   No trabeculæ and no cancellous bone. The bone is hard, the walls thin.

Structure. 1.   The external lamellae vary in number.   In some places there are three, in some six, in other places nine.   They are narrow, their lacunae are long, numerous and have very fine, long canaliculi.   Here and there canals traverse the entire thickness of them all.

2.   A ring of small, irregularly-shaped, incomplete Haversian systems.   They are best developed along the posterior ridge.   In the anterior wall they are few and confined to the region which is just below the surface.   Between the posterior and anterior middle lines they are rather indistinct, on account of their shape, arrangement and incomplete formation.   The canals are short or long, angular and branching.   They are very prominent and reunite with each other frequently.   Along some of them are lamellae, along others incomplete Haversian systems.   The entire ring impresses one as a transitional blending of lamellae into Haversian systems.   The lacunae are oval, numerous and have bushy canaliculi.

3.   The internal circumferential lamellae differ from others examined.   They form a thick, heavy ring around the medullary canal. There are thirteen to twenty of them, which, on the outer wall of the bone, merge into six laminae.   The whole ring of the internal lamellae forms about one-third of the thickness of the wall of the bone.   It is traversed by many canals extending from the medullary canal into the canals of the Haversian systems.   Their lacunae are long and numerous and their canaliculi are bushy.   The bone is of

the Haversian system type undeveloped. The peculiar feature of the bone is the thick, well-developed internal circumferential lamellae which merge into laminae.

Femur of the Eagle.
Pl. II, Fig. 11.

Antero-posterior diameter of the bone, 13 mm.; lateral, 14 mm. Antero-posterior diameter of the medullary canal, 11 mm., lateral, 11.5 mm.

Medullary canal is empty. Trabeculæ are present in the lower third. No cancellous bone. The bone is of medium hardness.

Structure. The external circumferential lamellae, six to ten in number, surround the bone, excepting at two posterior ridges, where they are interrupted by tendon attachments. They are fully developed. Their lacunae are long, narrow and concentrically arranged and their canaliculi are rather short and branching.

The central ring of bone is composed of six to twelve concentric laminae, interrupted here and there by poorly-developed Haversian systems. The canals which separate the laminae are relatively wide and, on account of their frequent communications with neighboring canals, they present the appearance of a coarse network.

The laminae are mostly short and consist of four to six or seven lamellae. Their lacunae are oval or round and their canaliculi are short and bushy. They indicate a low stage of development. A few poorly-developed Haversian systems are scattered among the laminae and in some instances appear to be circular dilatations of the concentric canals.

Internal circumferential lamellae, six to twelve in number, surround the medullary canal. They are fully developed and are frequently crossed by canals extending inward from the medullary canal. Their lacunae are long and narrow and their canaliculi are long and branching.

On the posterior surface are two ridges, one central and one on the posterior inner lateral border. The bone at these points consists of poorly-developed Haversian systems, separated by frequent wide canals, which pass to an apex at the outer surface of the ridges. The external circumferential lamellae are absent at these points and tendon insertions, interspersed by many canals, occupy the ridges.

The peculiar features of the bone are its undeveloped condition and its resemblance to the femurs of the peahen and turkey. The eagle's medullary canal is empty, like that of the peahen, but its laminar structure is more like that of the turkey. The type is incomplete laminar.

Femur of the Hawk.
Pl. II, Fig. 12.

Antero-posterior diameter of bone, 4 mm.; lateral, 4.3 mm.
Antero-posterior diameter of medullary canal, 3 mm.; lateral, 3.3 mm.
The medullary canal is empty. The bone is hard.

Structure. The section is surrounded by four to six external circumferential lamellae, fairly well developed. Their lacunae are more frequently oval than long. This fact indicates a less complete development. The canaliculi are bushy.

In the inner lateral posterior wall is a ridge to which are attached muscle tendons penetrating the external lamellae. Underneath the external circumferential lamellae is a thick ring of incomplete Haversian systems with oval lacunae and short, bushy canaliculi. The ring is crossed at all angles by wide, irregular canals, which are mostly confined to the ring. The Haversian systems are most prominent and best developed near the ridge of the lateral posterior region. This central Haversian system ring blends with the external circumferential lamellae.

The medullary canal is enclosed by six or eight well-developed internal circumferential lamellae with long lacunae and canaliculi. It is distinct from the central ring. No cancellous bone.

The peculiar feature of the bone is its low development. The type is incomplete Haversian system.

Femur of a Grouse.
Pl. II, Fig. 13.

Antero-posterior diameter of the bone, 5 mm.; lateral, 5.5 mm.
Antero-posterior diameter of the medullary canal, 4 mm.;
lateral, 5 mm.

The medullary canal is empty. Trabeculæ are present in the lower third. The bone is soft. No cancellous bone.

Structure. There are no distinct external circumferential lamellae. The bone, with the exception of a narrow ring of internal circumferential lamellae, is composed of short concentric laminae, separated by wide canals. Each lamina consists of two to four lamellae, with long, narrow or oval lacunae and long, branching or bushy canaliculi. The canals freely communicate with each other across the laminae. In the anterior wall (middle portion) is a slight prominence or ridge, consisting of three or four poorly-developed Haversian systems, situated close to the external surface and several whorls of lamellae arranged around short canals running in different directions. In the posterior wall are two ridges separated by a concave surface of bone. A single, poorly-developed Haversian system is found at the apex of each ridge, around which are collections of oval lacunae, with short, bushy canaliculi. Close to the internal circumferential lamellae are a few Haversian systems of a crude type.

Internal circumferential lamellae, three or four in number, surround the medullary canal. Their lacunae are long and narrow. The type of bone is the laminar.

The peculiar feature is the absence of complete Haversian systems. The bone resembles that of a turkey.

Femur of the Domestic Chicken.
Pl. III, Fig. 14.

Antero-posterior diameter of the bone, 9 mm.; lateral, 9 mm.
Antero-posterior diameter of the medullary canal, 7 mm.;
lateral, 7 mm.

The bone is practically round. The medullary canal is full. No trabeculæ and no cancellous bone. It is of medium hardness.

Structure. 1. Well-marked external circumferential lamellae, three to five in number, with long, narrow lacunae and branching canaliculi. They are quite distinct from the remainder of the bone.

2. A wide ring of irregularly-shaped, incompletely-developed Haversian systems. Some of the systems are circular in cross section, but most of them exhibit no definite shape. They run in various directions, are better developed on the outer side than on the inner side and tend to centralize at the posterior ridge and anterior surface. At the posterior ridge they occupy the entire thickness of the wall of the bone as far as the internal circumferential lamellae. At the ridge surface canals and bone lamellae take a direction at right angles to the long axis of the shaft. Along the anterior surface the systems are better developed and extend a short distance on both sides of the middle line. Interspersed between the systems are short lamellae. The lacunae of the systems are oval and the canaliculi are short and bushy. The Haversian canals are prominent in some places where they form a network, while in other places they form a parallel system.

3. Internal circumferential lamellae, four or five in number, completely surrounding the medullary canal. Their lacunae are long and narrow and their canaliculi are bushy.

Both external and internal lamellae are well-developed and distinct from the Haversian ring. The bone is of the Haversian system type undeveloped.

Femur of the Prairie Chicken.
Pl. III, Fig. 15.

Antero-posterior diameter of the bone, 5 mm.; lateral, 6 mm.

Antero-posterior diameter of the medullary canal, 4 mm.; lateral, 4.5 mm.

The medullary canal is empty. Trabeculae are present. The bone is soft. No cancellous bone.

Structure. There are no distinct external circumferential lamellae. The bone is composed of crude lamellae, crossed at all angles by short canals, some of which extend inward from the external surface. In the posterior and outer walls they unite and form a coarse network, while in the anterior and inner walls they

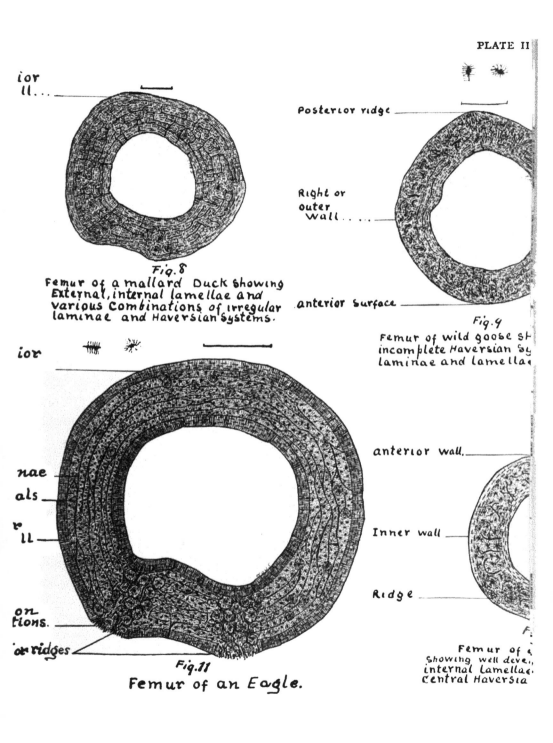

PLATE II

ior
ll...

Fig. 8
Femur of a mallard Duck showing
External, internal lamellae and
various Combinations of irregular
laminae and Haversian systems.

Posterior ridge

Right or
outer
Wall....

anterior Surface

Fig. 9
Femur of wild goose sh
incomplete Haversian sy
laminae and lamellae

anterior wall.

Inner wall

Ridge

Femur of a
Showing well deve
internal lamellae
Central Haversia

ior

nae
als
r
ll

on tions.
or ridges

Fig. 11
Femur of an Eagle.

do not. The lamellae are not distinct, but are blended together. Their lacunae are oval or narrow and their canaliculi are bushy or long and branching.

A very few crude Haversian systems are found interrupting the lamellae of the anterior and inner walls. They have only three or four lacunae around a minute canal. Their canaliculi are few. In the posterior wall are two ridges separated by a concave surface of bone. Two or three undeveloped Haversian systems are found in each ridge.

The internal circumferential lamellae, four to six in number, surround the medullary canal. They are well developed. Their lacunae are long and narrow and their canaliculi are long and branching. The type of bone is lamellar.

The peculiar features are the general absence of Haversian systems and the crude lamellar formation. The bone is not as far advanced as that of the grouse (Blue).

Femurs of the Domestic and Wild Turkey.
Pl. III, Figs. 16, 17.

Domestic turkey.   Antero-posterior diameter, 15 mm.; lateral, 17.5 mm.
Antero-posterior diameter of medullary canal, 10.5 mm.; lateral, 13 mm.

Wild turkey.   Antero-posterior diameter, 9 mm.; lateral, 11 mm.
Antero-posterior diameter of medullary canal, 7 mm.; lateral, 8 mm.

Since the two bones resemble each other closely, one description will answer for both. The medullary canals are full. No trabeculæ; no cancellous bone. The type of structure is laminar. The bones are soft. The medullary canals are relatively large and the walls of the bones are thin.

Structure. Around the outside are four or five external circumferential lamellae, between which are long, narrow lacunae, with many canaliculi. Along the posterior ridges of the two femurs are small areas of incomplete, irregularly-shaped Haversian systems which occupy nearly the entire thickness of the posterior walls of the bones. They extend nearly to the outer surfaces, where they seem to be projected in a network of bone extensions, which pass to the surfaces of the posterior ridges, where they blend with the tendon attachments. In the anterior walls are small areas of the same irregular Haversian systems. The systems do not reach either surface. In both of these regions the systems do not appear to run parallel with the outer and inner surfaces of the bones, but extend in different directions. Since their cross sections are circular, elliptical, angular and very irregular in shape, no one direction could be followed by all. The Haversian canals are large, the lacunae are oval and their canaliculi are numerous and bushy.

Between the posterior ridges and middle anterior surfaces, and constituting eight-tenths of the whole section of the bones, are concentric laminae, fifteen to eighteen or twenty in number, separated by prominent canals and crossed at frequent intervals by smaller canals extending from both surfaces of the bones. The laminae are composed of three or four lamellae, between which are oval lacunae, with short, bushy canaliculi. Here and there a lamina is interrupted by a Haversian system.

Around the medullary canal the internal circumferential lamellae are not distinct from the adjoining laminae. The peculiar feature of the turkey femurs is the laminar formation.

It may be noticed that the Haversian systems present in the section are found in the posterior walls and ridges and in the anterior walls, where muscular stress is greatest.

Femur of the Peahen.
Pl. III, Fig. 18.

Antero-posterior diameter of the bone, 10 mm.; lateral, 11 mm.

Antero-posterior diameter of medullary canal, 8 mm.; lateral, 10 mm. Bone is thin; medullary canal is large, empty; has a network of trabeculæ which extends from one wall in a downward direction to the opposite wall. They are most numerous near the extremities of the bone. No cancellous bone. The bone is extremely hard.

Structure. The usual three divisions of the bone into external circumferential lamellae, middle ring of Haversian systems and internal circumferential lamellae do not appear. The bone consists of a network of canals separating short, concentric laminae. The canals intersect at all angles. They are wide. The laminae are composed of four or five lamellae with oval lacunae and relatively few rather short, bushy canaliculi. In the vicinity of the posterior ridge a few incomplete Haversian systems are found. In other parts of the section, here and there, a system appears without any apparent signification. The bone is peculiar in the absence of Haversian systems. The bone trabeculae are composed of three or four lamellae with long, narrow lacunae and branching canaliculi. A few systems are present. The bone is of the laminar type, but does not show a development equal to that of the turkey. However, it belongs to the same structural type and can be recognized as such with no difficulty.

Femur of a Yellow Hammer.
Pl. III, Fig. 19.

Antero-posterior diameter of the bone, 2.5 mm.; lateral, 3 mm.
Antero-posterior and lateral diameters of the medullary canal, 0.5 mm.
The medullary canal is full and situated close to the posterior wall. The bone is soft.

The femur consists of a wall of compact bone, composed of three or four external circumferential lamellae and about the same number of internal circumferential lamellae, between which are a few irregularly-developed Haversian systems. Large canals extend transversely across the walls of the bone communicating with the meshes of the central bone structure.

The central portion of the bone usually occupied by the medullary canal is composed of a fine cancellous bone formation, with the exception of a fine medullary canal about the size of a fine sewing needle, situated close to the posterior wall. The femur is therefore nearly solid bone. The cancellous center is composed of fine lamellae forming a meshwork extended from the internal circumferential lamellae. The meshes are filled with granular material, insoluble in ether or chloroform. It is difficult to grind out this material and leave uninjured the meshwork. The lacunae are small, round or oval and their canaliculi are short, bushy and infrequent. Round and oval lacunae appear to be antecedent forms of long, narrow lacunae. In undeveloped Haversian systems the lacunae are round or oval, while in the complete systems they are long and narrow. It is possible that the pressure of development accounts for the variation in shape—the medullary canal is full and extremely small. Its position is unlike that of other canals.

The peculiar features of this femur is the central bone formation and small eccentric medullary canal. Although the yellow hammer is a good flyer, its femur is practically a solid bone. The type is lamellar-cancellous.

Femur of the Crow.
Pl. IV, Fig. 20.

Antero-posterior diameter of the bone, 4 mm.; lateral, 3 mm.

Antero-posterior diameter of medullary canal, 2.5 mm.; lateral, 2 mm.

The medullary canal is full. No trabeculæ and no cancellous bone. The bone is soft.

Structure. 1. External circumferential lamellae, six or eight in number, form a wide lamina around the bone. The lamellae are not clearly defined, the lacunae are oval, with bushy, connecting canaliculi. A little to the outer side of the anterior mid-line, the lamina dips down and forms a semi-circular depression about ¼ mm. in diameter.

2. A central wide ring of transverse canals, enclosed by incomplete Haversian systems and irregular lamellae. Their lacunae and canaliculi are the same as above described. There is very little difference in the structure of the various parts of the bone. Possibly the Haversian systems are a little better developed in the posterior wall.

3. Internal circumferential lamellae, two or three in number, form a narrow ring around the medullary canal. Their lacunae are narrow and long and their canaliculi are long and branching.

The bone is of the mixed type. The peculiar features are the external lamina, uniform central ring and semi-circular depression.

PLATE III

erior
all....

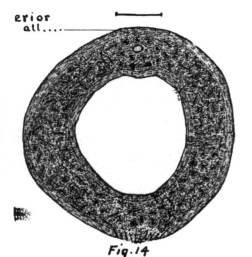

Fig.14

emur of domestic chicken
howing external, internal lamellae
nd irregular Haversian Systems.

Anterior wall

Inner wall

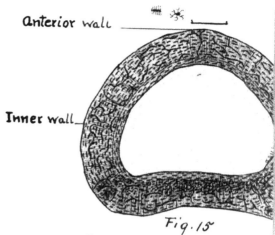

Fig.15

Femur of a Prairie Chicke
showing lamellar type.

ular
'sian
ms of
rior
e....

n to
een
nae.

ular
rsian
ms of
rior
e....

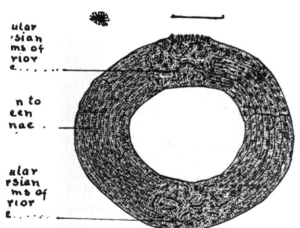

Fig.17

Section of the femur of a wild
turkey showing laminae and
irregular Haversian Systems on
anterior and posterior ridges.

Posterior ridge

incomplete
Haversian Systems

Haversian canals

Laminae

Trabecula...

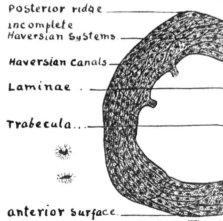

anterior surface

Fig.18

Femur of a peahen she
of Haversian Canals, SI
Haversian Systems an

Femur of a Blue Jay.
Pl. IV, Fig. 21.

Antero-posterior diameter of bone, 2.5 mm.; lateral, 2.5 mm.
Antero-posterior diameter of medullary canal, 2.5 mm.; lateral, 1.5 mm.
The bone is nearly round. The medullary canal is full. The bone is hard.

Structure. The anterior and outer walls are composed of four or five incompletely-formed laminae, frequently uniting. The laminae consist of three or four lamellae. The inner and posterior walls are composed of lamellae interrupted by a few incompletely-developed Haversian systems. The lacunae are oval. The canaliculi are few and short.

The type of bone is lamellar. The peculiar feature of the bone is its close conformity to the.lamellar type and absence of Haversian systems.

Femur of the Opossum.
Pl. IV, Fig. 23.

Antero-posterior diameter of the bone, 7 mm.; lateral, 8.5 mm.
Antero-posterior diameter of the medullary canal, 3.5 mm.; lateral, 5 mm.
Medullary canal is full. No trabeculæ and no cancellous bone. The bone is soft.

Structure. The bone presents a rudimentary appearance. It is composed of two wide external lamellar bands of incomplete formation separated by a very narrow band of imperfectly-developed Haversian systems, the whole occupying two-thirds of the posterior, outer and anterior walls. The lamellar bands simply give the general appearance of lamellae, but are really composed of large oval lacunae, with extensive, bushy canaliculi forming an intricate network. At short intervals short transverse canals appear, with many radiating canaliculi. Just internal to this lamellar band is a narrow crescent of very incomplete Haversian systems occupying the anterior, outer and posterior walls. The systems are merely canals, from which radiate numerous straight canaliculi, with a few oval lacunae around their apparent circular boundaries. Around the medullary canal of the anterior, outer and posterior walls internal circumferential lamellae are well developed, reaching their greatest thickness in the outer wall. Their lacunae are long and narrow and their canaliculi are long, straight and branching.

The inner wall of the bone is extended in the form of a heavy ridge. It is composed of heavy, oblique canals, from which are sent off dense networks of large canaliculi. This peculiar arrangement forms the external half of the ridge. The internal half consists of incomplete Haversian systems, arranged in oblique rows, converging to a central point in the middle of the ridge. The systems are similar to those described above. No internal circumferential lamellae are found in this region. The bone is very poorly developed and belongs to a type of structure characterized by incomplete development.

The peculiar features are the undeveloped lamellar and Haversian systems and the heavy oblique canals. The only complete structure is the partial internal circumferential ring of lamellae.

Femur of the Musk Rat.
Pl. IV, Fig. 24.

Antero-posterior diameter of the bone, 5 mm.; lateral, 6.5 mm.
Antero-posterior diameter of medullary canal, 2.5 mm.; lateral, 3 mm.
Medullary canal has no contents. From the medullary surface of the inner wall project a few fine trabeculae. No cancellous bone. The bone is soft.

Structure. The bone is peculiar in many respects. The inner wall of the bone is extended in the form of a ridge, which is composed of a network of laminae and canals running transversely from above downwards and occupying the outer four-fifths of the ridge. Each lamina consists of two or three lamellae, with long or oval lacunae and long branching or bushy canaliculi. The inner one-fifth of the ridge wall consists of a network of laminae running from the medullary canal to the outer network. Within its meshes are found canals from which radiate many long, branching canaliculi. The remainder of the bone (anterior, outer and posterior walls) is composed of a very irregular, wide internal ring of lamellae sur-